SCIENCE TEACHER

DON'T MAKE ME USE MY SCIENCE TEACHER VOICE

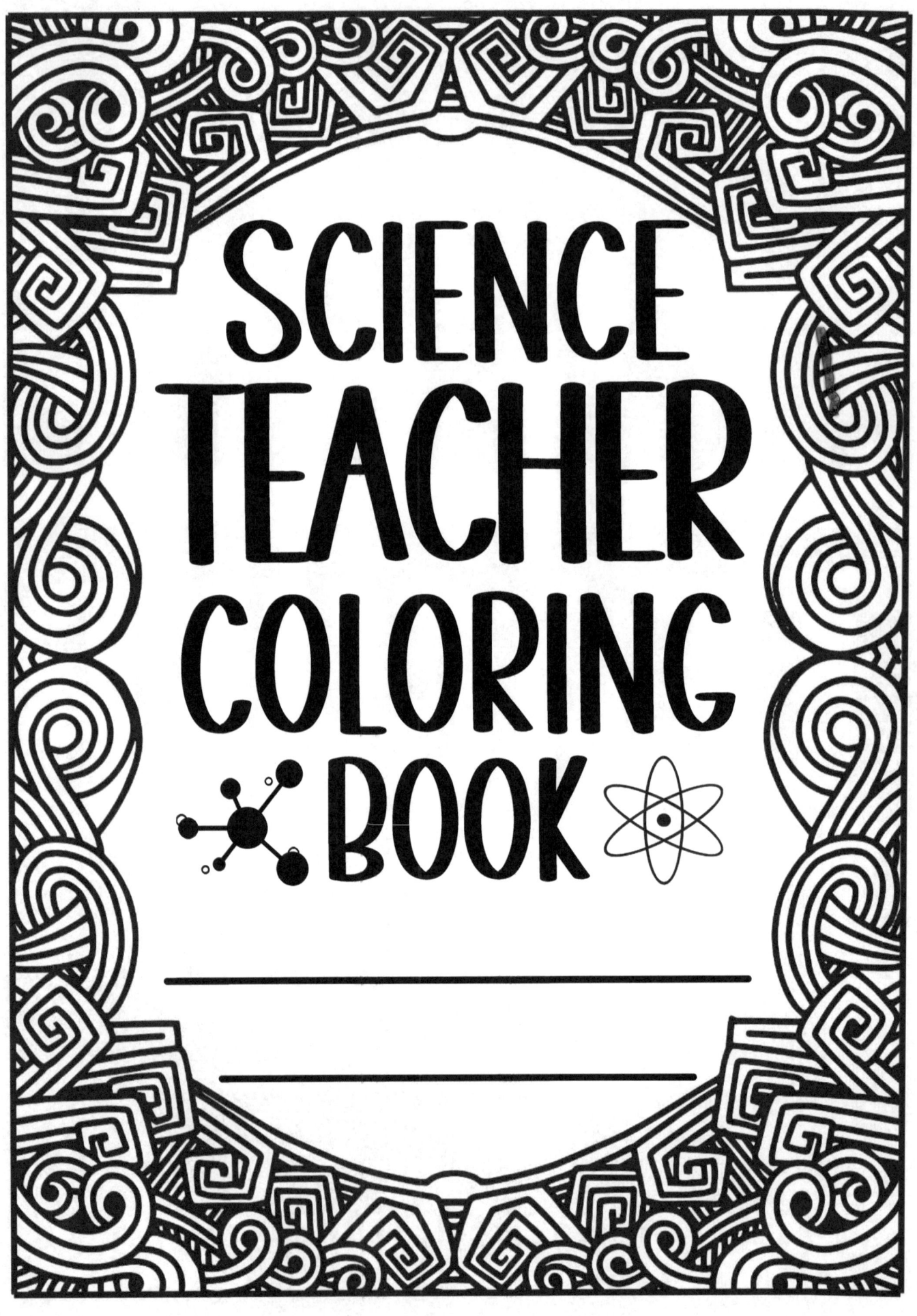

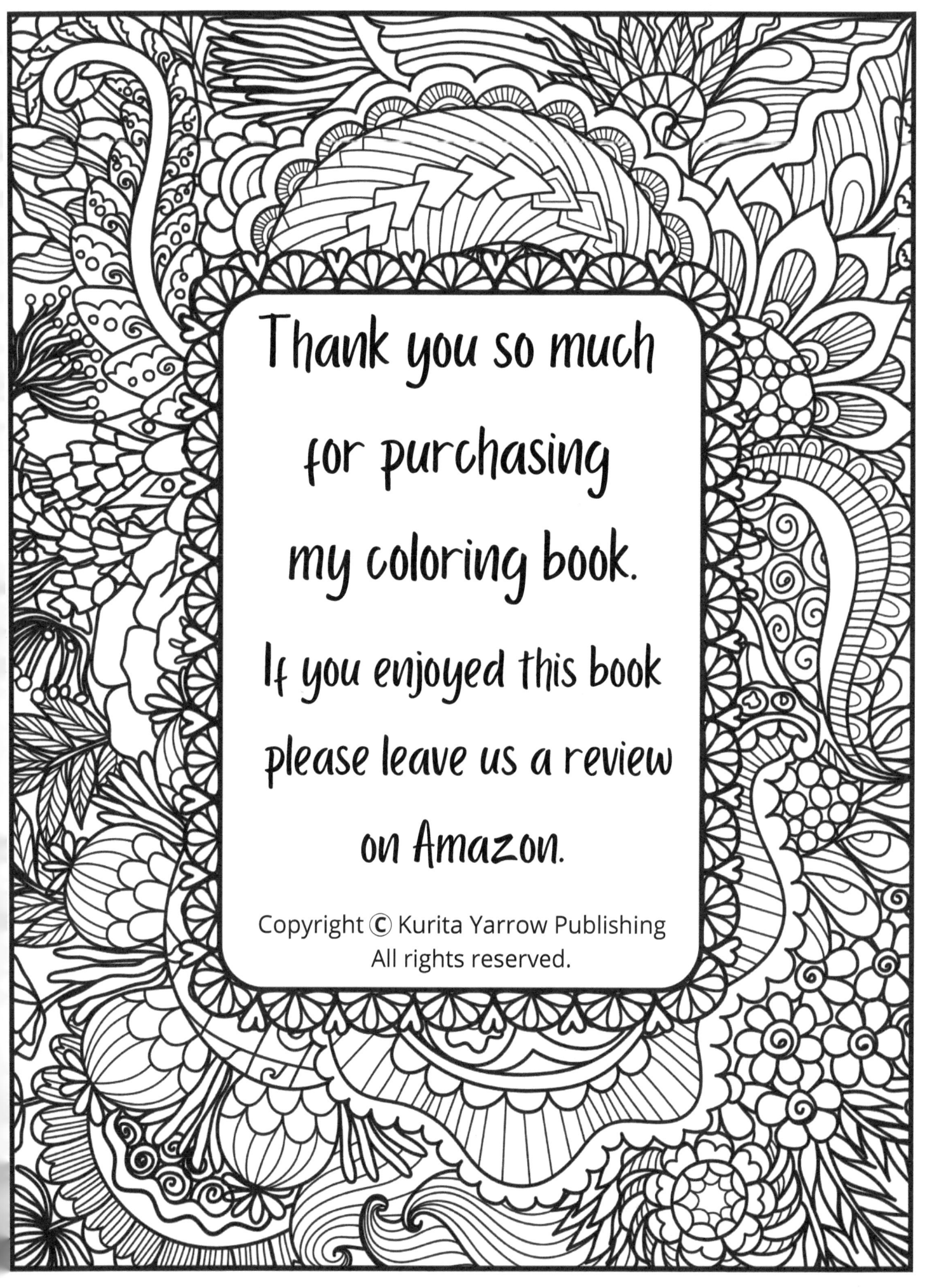

Thank you so much
for purchasing
my coloring book.

If you enjoyed this book
please leave us a review
on Amazon.

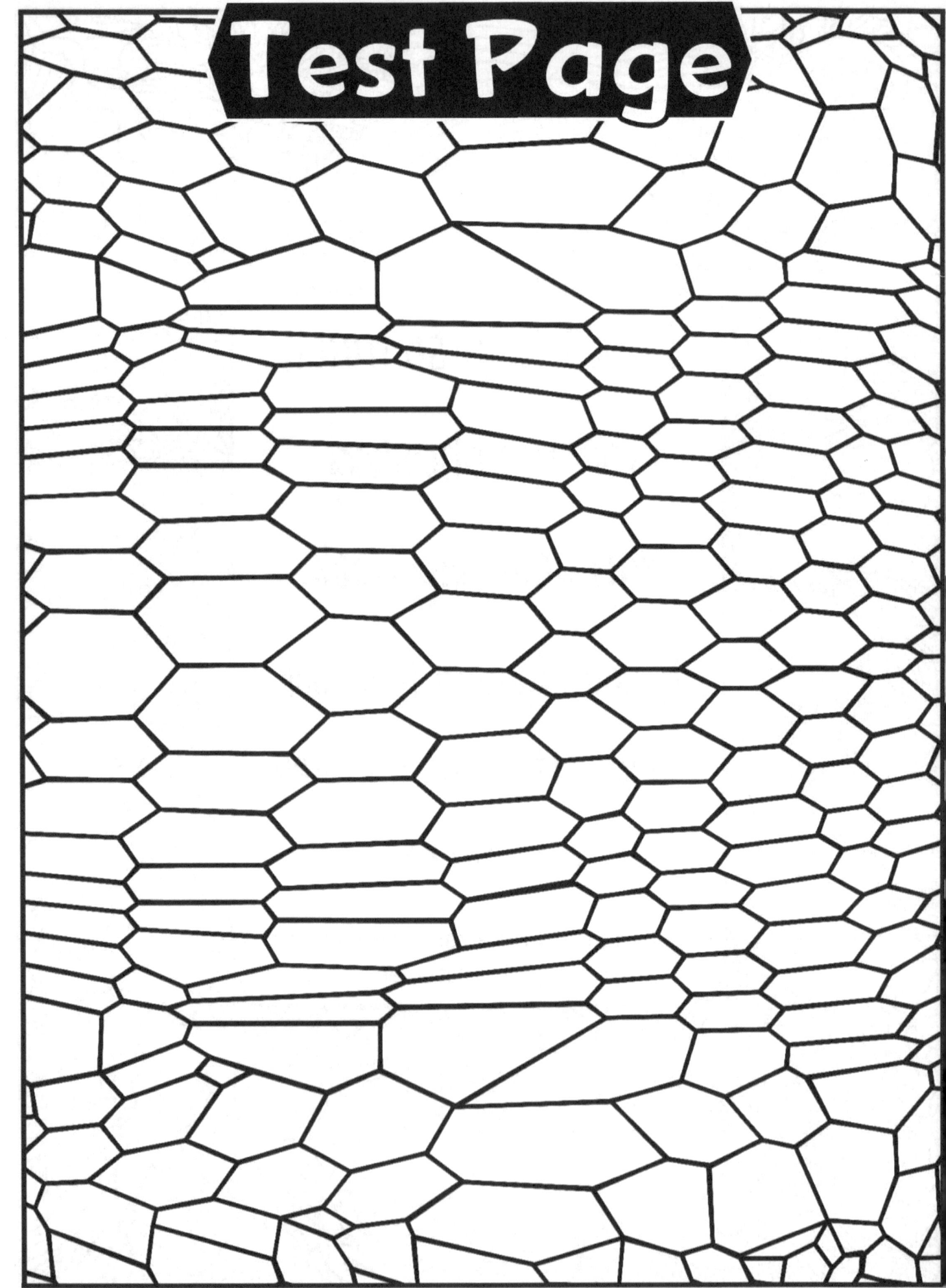

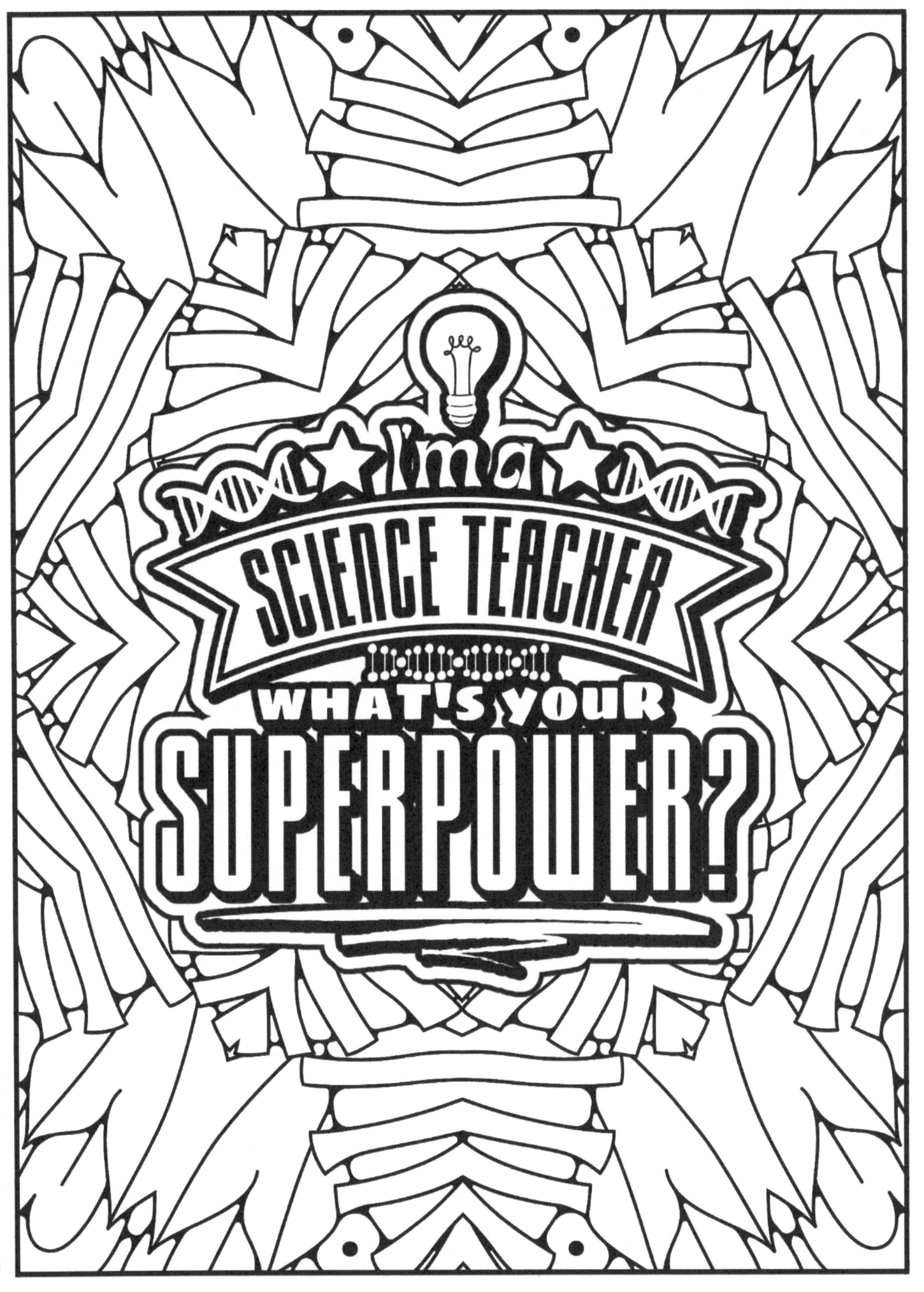

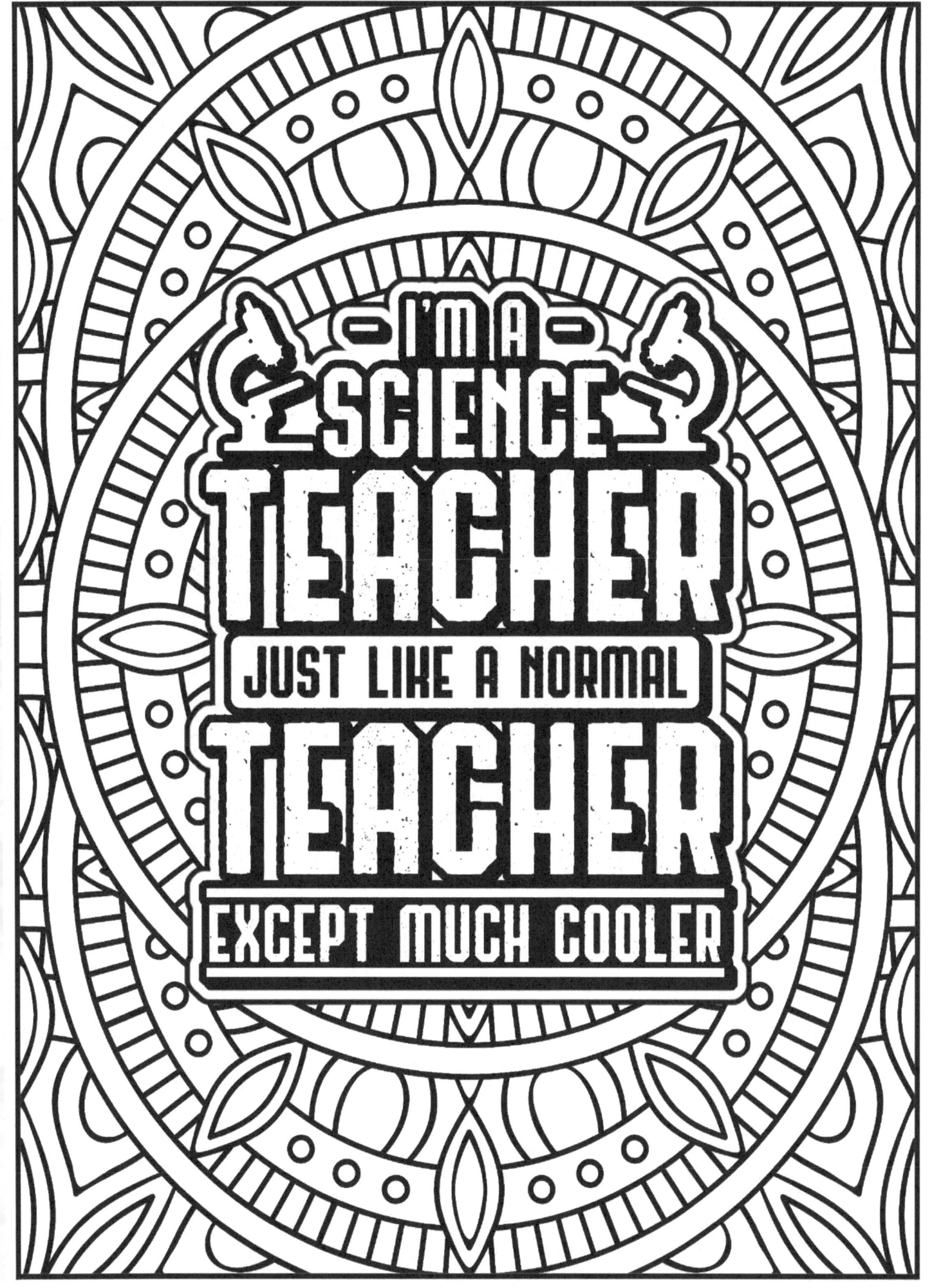

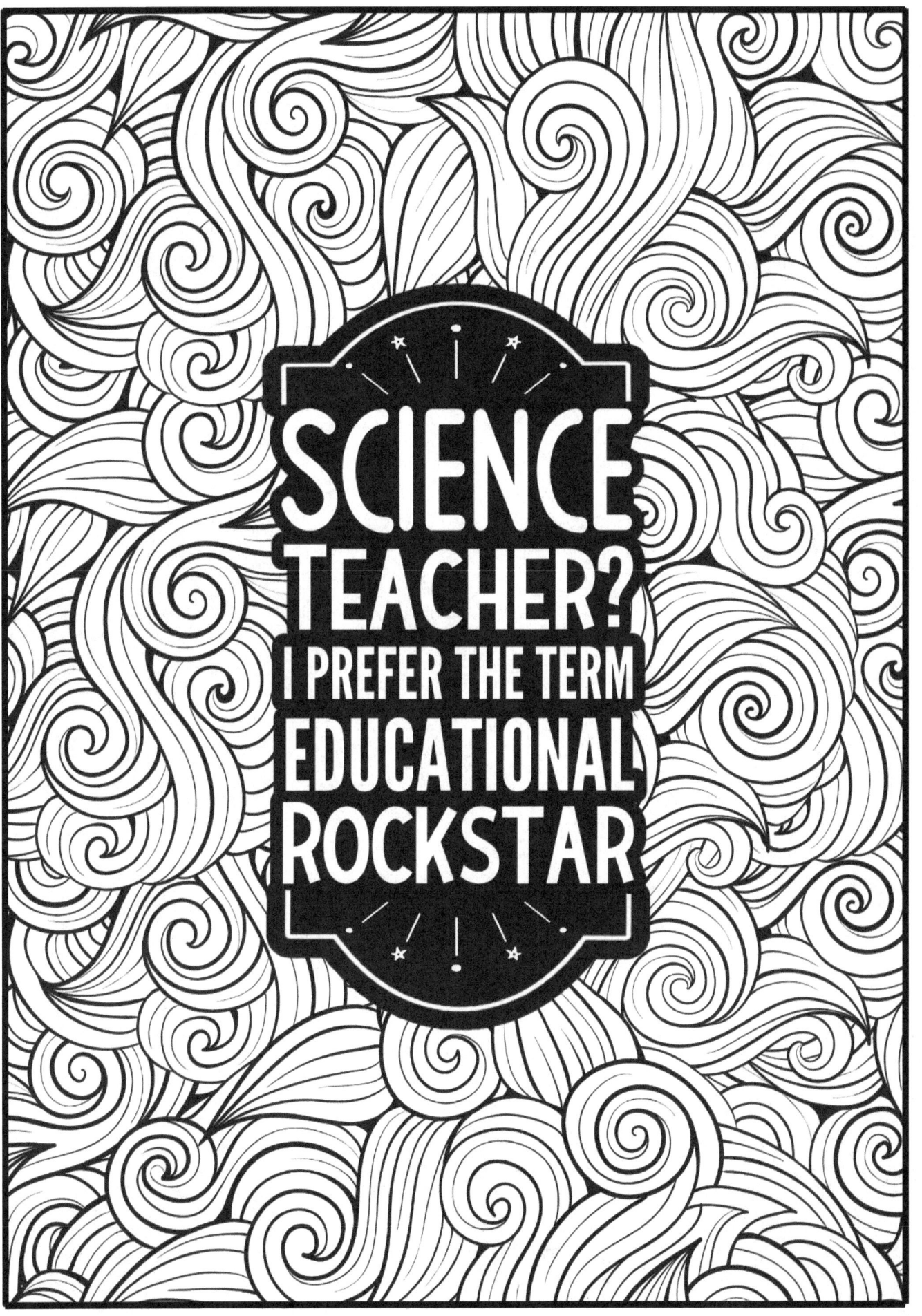

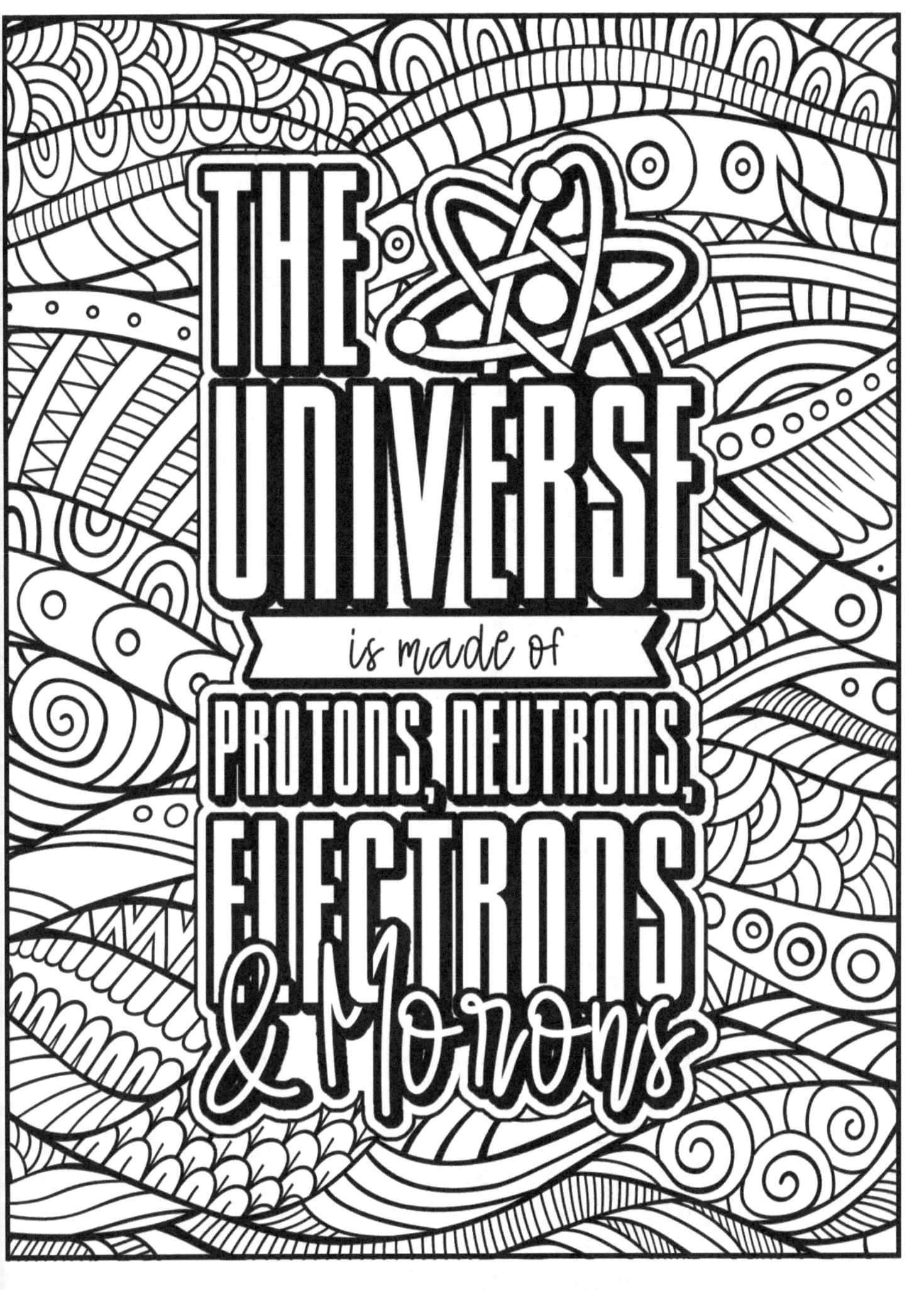

123

Ah!

(230)

**The element
of surprise**

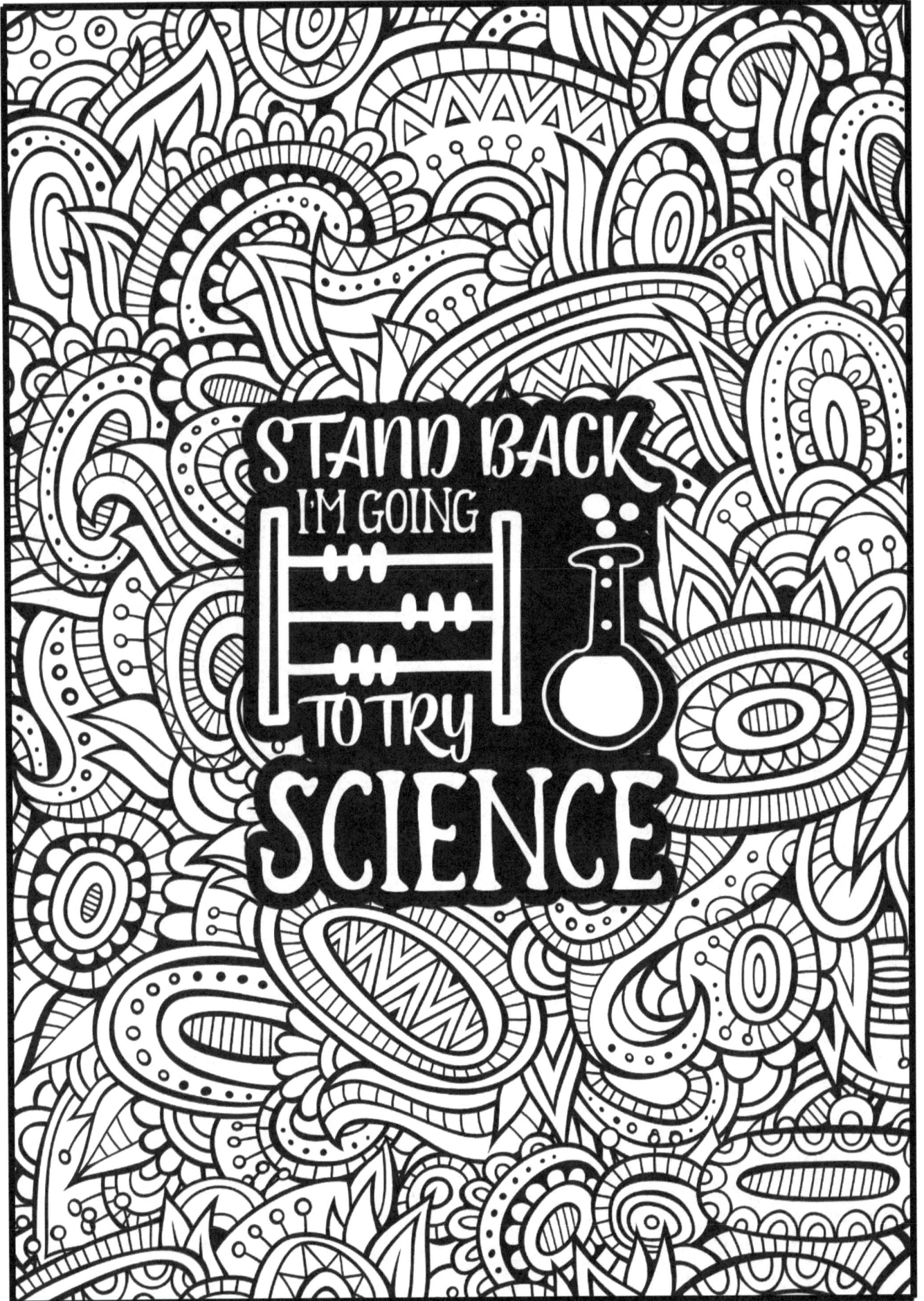

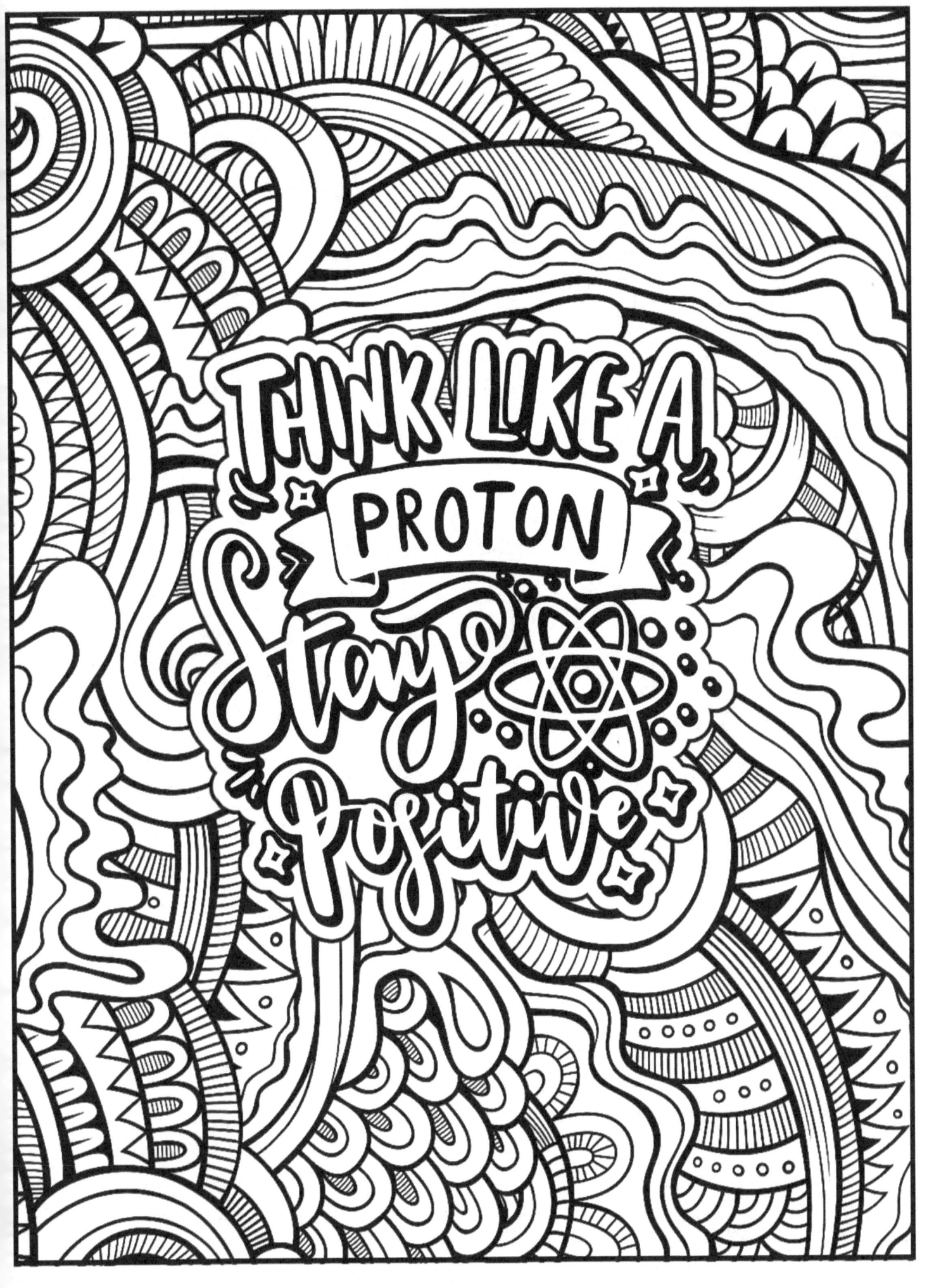

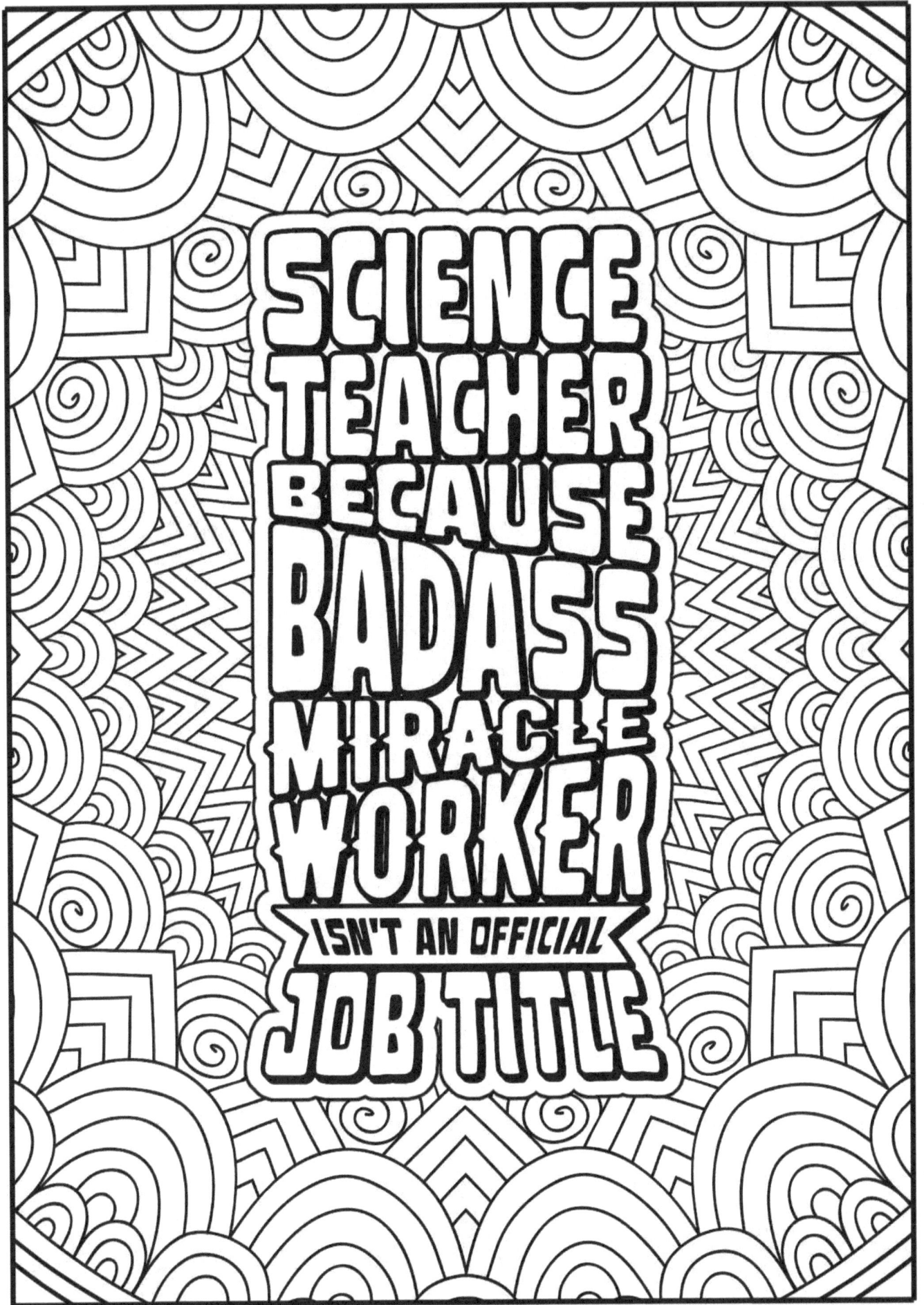

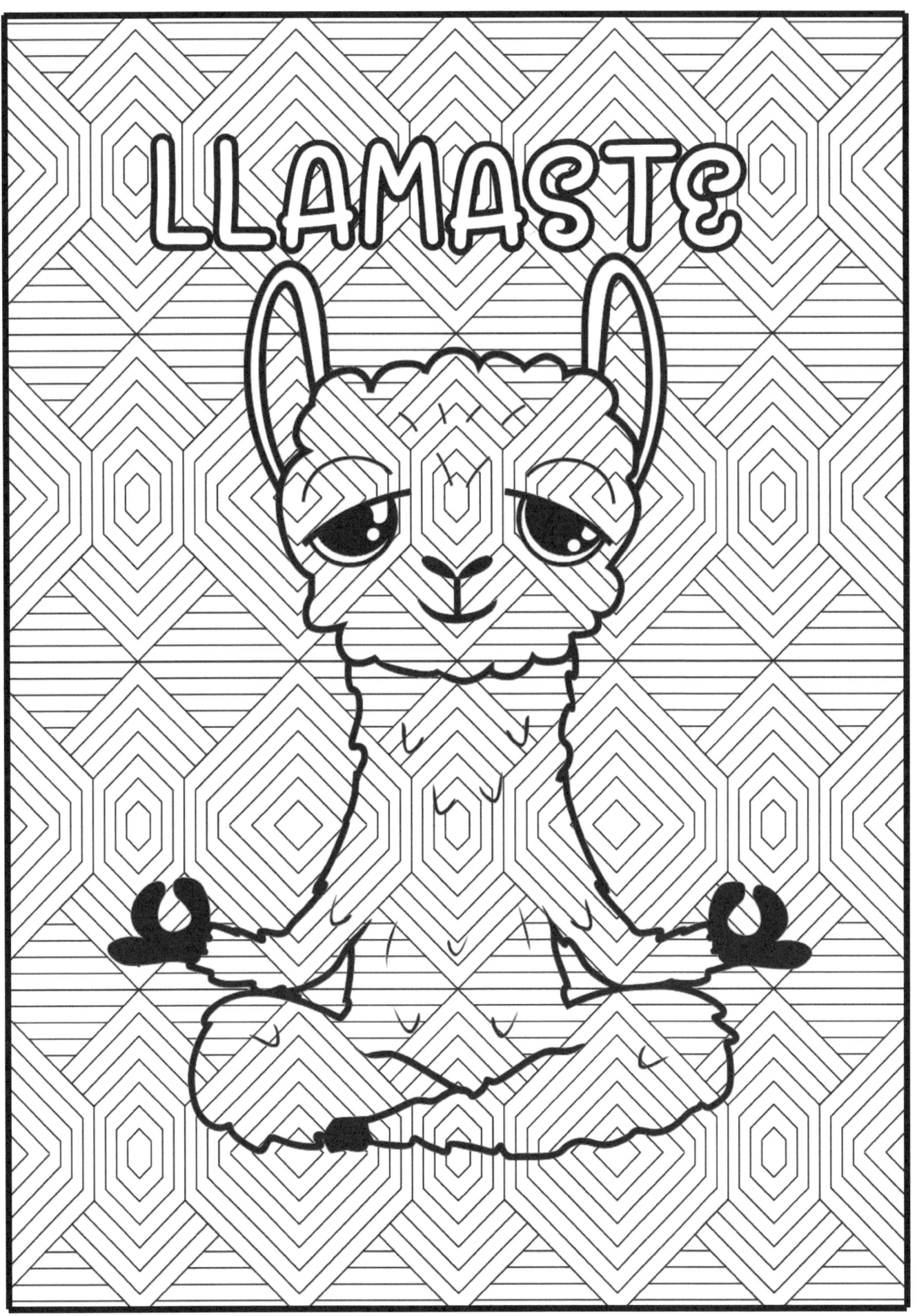

Made in United States
Troutdale, OR
03/03/2024